STEAM & Me™
ARTIFICIAL INTELLIGENCE

DINAH WILLIAMS

Starry Forest Books

SCIENCE · TECHNOLOGY · ENGINEERING · ARTS · MATHEMATICS

Draw a super-smart robot. Create your own wind energy. Find out if your teeth are as sharp as a shark's. Go back in time to the world of dinosaurs or rocket into space. Power up that scientific brain of yours with STEAM & Me!

Photos, facts, and fun hands-on activities fill every book. Explore and expand your world with science, technology, engineering, arts, and math.

STEAM&Me is all about you!

Great photos to help you get the picture

New ideas sure to change how you see your world

A.I. programs can help write songs. If a musician has an A.I. program listen to a bunch of songs, that program can create a new song that sounds like the others.

Do you hear what I hear?

Computers don't have ears. But they can listen to songs through microphones. They hear the different sounds in a song, like a drumbeat or singing voices. They can also tell if the song is fast or slow. Those skills help an A.I. program help you. If an A.I. program knows your favorite song, it can find similar songs for you to listen to.

STEAM&Me
Try humming these three songs: "Twinkle, Twinkle, Little Star," "Baa, Baa, Black Sheep," and "ABC." What similarities do you notice? Now you're thinking like an A.I. program!

See What You Like!
A.I. TV services use information about the shows you watch to come up with ideas for new shows you might like.

Hands-on activities to spark your imagination

Fascinating facts to fill and thrill your brain

Brain games!

How do you remember the words to your favorite song? Or make your hands move to catch a ball? Or learn the names of all the colors? Your brain does it! It is always working hard and learning new things. Scientists wondered: What if we could make machines that work like a human brain? So they started trying to teach computers to do things brains do. And it worked! When a machine learns to act like a brain, it's using what's called artificial intelligence, or A.I.

Think fast! In boxes like these throughout the book, you will find activities and experiments to help you learn how A.I. works—and how to think like a computer yourself.

Play Ball!

For a kid, learning to throw, catch, or hit a ball takes practice, and lots of brain power. Scientists can teach robots to play baseball by training them to practice and learn as we do.

What is artificial?

Some things that seem real or natural are actually made by people. That makes them **artificial**. Some candies have artificial flavors to make them taste like real fruit. Something that has artificial intelligence has intelligence that's made by humans.

Some fruit candies have real fruit juice in them, but many use artificial flavors. A flavorist mixes different chemicals together until the mixture tastes just like a sweet, fruity treat.

STEAM Me

When have you seen something artificial? Have you ever played with toy food? Or seen a sports field with fake grass, or turf? How could you tell that it wasn't real? Make a list of which senses you can use to tell if something is real or artificial.

Let It Snow!

When it doesn't snow enough, ski resorts make artificial snow. They use huge machines that shoot frozen water into the sky. Pretty soon, it looks and feels like it's "snowing."

Where's the Smell?

Artificial flowers are made out of plastic or fabric, but they look like they were grown in a garden.

You need to learn to be smart.

Ouch! Have you ever been so excited to take a bite of something hot that you burned your mouth? Did you blow on the next bite to cool it down? That's your intelligence working. Learning from what has happened is one thing that makes our brains so **smart**, or intelligent.

Bird Brain

Birds learn, too. Crows are smart enough to solve problems. Crows can use tools like sticks to reach tasty bugs in places too narrow for their beaks.

A baby's brain has a lot to learn. When you play peekaboo with a baby and hide behind your hands, she believes you disappear. It takes a lot of playtime before she learns you're still there.

Sophia is a robot with A.I. It can show feelings like happiness or sadness.

When computers are smart, we say they have artificial intelligence.

Your brain learns on its own. But computers need human help to learn new things. Scientists write **programs** with steps telling a computer what to do. These programs teach computers how to understand the world. Like the human brain, a computer with artificial intelligence, or A.I., gets smarter the more it works.

Let's Play!

The Deep Blue computer learned to play chess. It learned the game so well that in 1997 it was the first machine to beat a human world chess champion.

Practice, practice, practice.

Playing a musical instrument can be hard to do at first. But it gets easier with practice. Computers have to practice to get better at things, too. While learning to use a new language, a computer might practice by looking at the language's words over and over. Today, computers are taught to do some jobs that only people used to do. A computer's job might be talking to people in different languages or counting money for a bank.

Chores No More!

Some robot vacuums use A.I. to learn where furniture and other obstacles are in the rooms they vacuum.

Think like an A.I. program! Draw a Venn diagram, two circles that overlap in the middle. Now think about a tiny dog. In one circle, write things that make a tiny dog different from a big dog. In the other circle, write what makes the big dog different from the tiny dog. Use the middle section to list what the two dogs have in common.

What makes a dog a dog? To learn, a computer needs to look at hundreds of images to find out how a dog looks different from other animals, and also in what ways all dogs look alike.

Rubik's Cube Race

One A.I. computer can solve the Rubik's Cube in less than 1 second. That's about as long as it takes to say the word "abracadabra!"

Gooaall!

RoboCup is a soccer game played by robots. Scientists are trying to build a robot soccer team that will play the game as well as the best human team.

Checkers was one of the first games played by a computer. But it took almost **50 years** before a computer was able to beat the best human checkers player.

Who plays it best?

Do you like playing games? Computers can play them, too. Computers work by following a series of steps in a program. Game rules are like steps in a computer program. A.I. computers get good at games quickly because they can play again and again without stopping to rest or get a snack.

Scientists teach computers to play games by first writing a program that tells the computer all the rules. What's your favorite game? Teach it to a grown-up or a friend. Tell them the program, or the rules, then start playing!

Get face to face with a computer.

How do you recognize your friends at school? Easy! When you first met, you learned what they look like. Now you remember. Their eyes might be brown, their hair curly. **Similar** to how your brain creates maps of your friends' faces to use later, A.I. computers make maps of faces so they can recognize them, too. They try to figure out who someone is by looking at their nose, eyebrows, cheeks, mouth, and all kinds of other little details.

How might you trick an A.I. program so it can't recognize you? Look in a mirror and make a funny face. Try to make your face look as different as possible. Are there parts of your face you can't change?

STEAM & Me

MATCH FOUND

MATCH FOUND

MATCH FOUND

If a company is doing top-secret work, it might use computers that recognize faces to make sure only the right people can get into the building.

Same but Different

Some A.I. programs have trouble telling twins apart. Can you guess why? Twins can have similar faces. Some twins look exactly alike.

A.I. programs can help write songs. If a musician has an A.I. program listen to a bunch of songs, that program can create a new song that sounds like the others.

Try humming these three songs: "Twinkle, Twinkle, Little Star," "Baa, Baa, Black Sheep," and "ABC." What similarities do you notice? Now you're thinking like an A.I. program!

Do you hear what I hear?

Computers don't have ears. But they can listen to songs through microphones. They hear the different sounds in a song, like a drumbeat or singing voices. They can also tell if the song is fast or slow. Those skills help an A.I. program help you. If an A.I. program knows your favorite song, it can find similar songs for you to listen to.

See What You Like!

A.I. TV services use information about the shows you watch to come up with ideas for new shows you might like.

Ask me anything!

It might feel silly to say "hello" to a computer, but some computers have A.I. that can say "hi" back! So do some phones and other speakers. They listen to your voice and answer your questions. They learn from their mistakes. So the more you use them, the better they get at guessing what you want.

Ask for Help

Some phones or speakers have A.I. helpers called *digital assistants*. An assistant is someone who helps you, and digital means it's electronic.

A smart speaker is always listening for you. But you need to say a special word, like its name, so it knows to pay attention to any questions.

A.I. computers are great at learning languages, which have rules they can follow. If you take a trip to another country, you can use an A.I. program that listens to what other people say and translates it into your language.

You can have a conversation with a computer.

Some A.I. computers are so good at acting human that they can trick us into thinking they are real people. A special kind of computer program called a *chatbot* talks to people by typing words onto a computer while you type back. Some chatbots have A.I. Someone might have a whole conversation with an A.I. chatbot and not realize that the "person" writing back to them isn't real!

The Write Stuff

Some A.I. programs turn spoken words into written words. Just tell a story and the computer will type it up for you!

Beep, beep! A.I. drivers on the road.

Cars didn't always have seat belts. Now all cars have them to keep people safer. A.I. programs in cars are helping people drive better and be safer now, too. Some can beep to let you know another car is too close to you. Others might show you a different route if you get stuck in traffic.

A.I. cars are called *smart cars*, and some are so smart, they can drive themselves.

This car has no steering wheel, brake, or gas pedal. But it can still take you where you need to go because it has an A.I. program!

Where's the Driver?

In some cities in China, self-driving buses take people all over town.

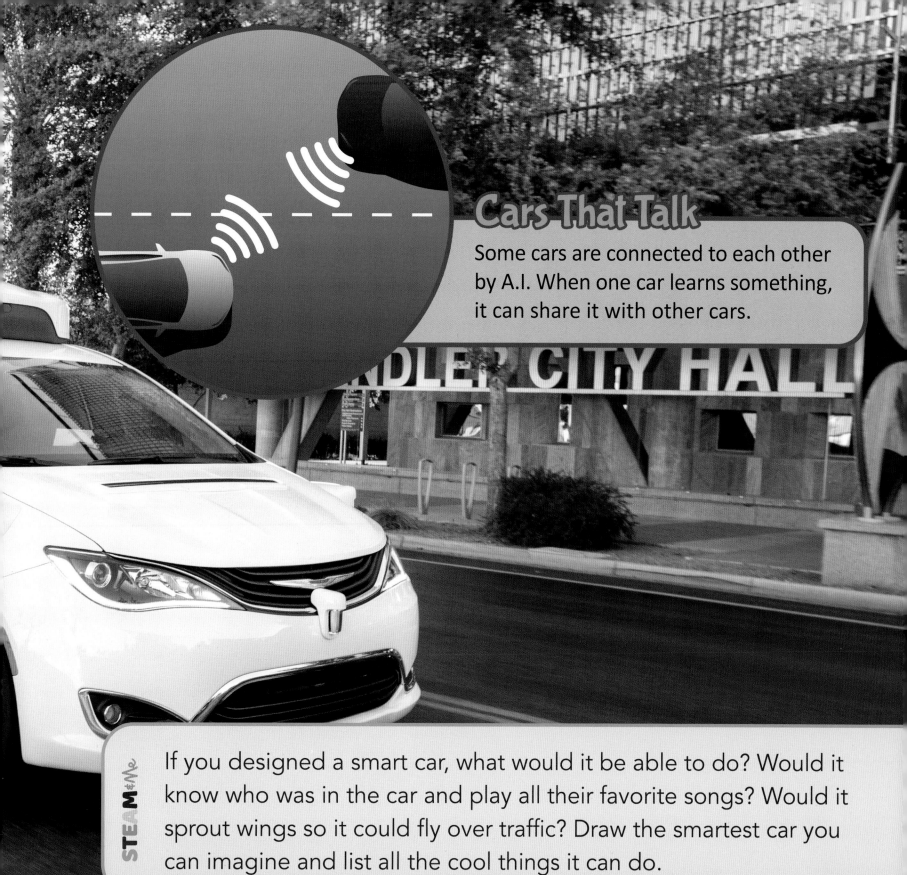

Cars That Talk

Some cars are connected to each other by A.I. When one car learns something, it can share it with other cars.

STEAM If you designed a smart car, what would it be able to do? Would it know who was in the car and play all their favorite songs? Would it sprout wings so it could fly over traffic? Draw the smartest car you can imagine and list all the cool things it can do.

Aibo is a robo-pup that is trained using A.I. Like a puppy, it doesn't always do what it's told. As it gets to know you, it gets better at doing tricks.

Let's play!

Is it a toy? Is it a pet? Smart toys are kind of like both put together. Like a pet, a smart toy can cuddle with you when you need a hug and learn to play with you if you teach it how. Some smart toys can even hear your voice and learn to listen to you. But you don't have to feed them or take them for a walk—unless you want to!

What Did You Say?

The Furby was one of the first learning toys. It came out of the box speaking only Furbish, but you could teach it some words in English.

Digital doctors are here to help.

To help when we're sick, doctors must know a lot about our bodies. Sometimes they use A.I. to **diagnose** problems. It's like having a super-fast, extra-smart brain working with them. A.I. can look at information about people who are sick and spot clues that help doctors discover better ways to treat their illnesses.

If your doctor thinks you might have broken a bone, she'll take an image called an X-ray. A.I. programs have learned by looking at thousands of other people's X-rays and can help your doctor understand yours.

Doctor Watson

A computer called Watson was built to answer difficult game show questions. Now it helps doctors learn about patients and pick out the best medicines for them.

Getting smarter all the time!

The better humans get at teaching A.I. programs how to act like a human brain, the better they will get at "thinking" like us.

What is your favorite thing that A.I. can do? If you could play any game with an A.I. computer, what would it be? If you could program a computer to learn anything, what would you teach it? Try writing a program that would tell the computer what to do!

Glossary

Learn these key words and make them your own!

artificial: made by people, sometimes trying to copy something from nature. *Some sports are played on* artificial *grass, called turf.*

diagnose: figure out what is causing a problem, like when someone's sick. *Your doctor can* diagnose *the flu and help you feel better.*

program: a set of steps written by people, telling a computer what to do. *A computer scientist writes a* program *that teaches a computer to play checkers.*

similar: having things in common with something else. *Your eyes are a* similar *color to my eyes.*

smart: able to learn or show good judgment. *Looking both ways before crossing the street is a* smart *decision.*

X-ray: a special kind of photograph that shows what's inside a person's body. *A doctor takes an* X-ray *to see if you have a broken bone.*

STEAM & Me and Starry Forest® are trademarks or registered trademarks of Starry Forest Books, Inc. • Text and Illustrations © 2020 and 2021 by Starry Forest Books, Inc. • This 2021 edition published by Starry Forest Books, Inc. • P.O. Box 1797, 217 East 70th Street, New York, NY 10021 • All rights reserved. No part of this publication may be reproduced, stored in a retrieval system, or transmitted in any form or by any means (including electronic, mechanical, photocopying, recording, or otherwise) without prior written permission from the publisher. • ISBN 978-1-946260-87-1 • Manufactured in China • Lot #: 2 4 6 8 10 9 7 5 3 1 • 03/21

ASP: Alamy Stock Photo; IS: iStock; SS: Shutterstock. Cover, LightField Studios/SS; 5, Suzanne Tucker/SS; 5, (LO) Itsuo Inouye/AP/SS; 6-7, Evlakhov Valeriy/SS; 7, (UP) Vova Shevchuk/SS; 7, (LO) Phichai/SS; 8, Auscape International Pty Ltd/ASP; 8-9, Creativa Images/SS; 10, paparazzza/SS; 11, Andrey_Popov/SS; 12, esp2k/SS; 13, Koldunov/SS; 14, (UP) David Gilder/SS; 14, (LO) FeelGoodLuck/SS; 14-15, bizoo_n/IS; 17, Zapp2Photo/SS; 17, (LO) Lopolo/SS; 18-19, Iaremenko Sergii/SS; 19, Prostock-studio/SS; 20, Zapp2Photo/SS; 21, Andrey_Popov/SS; 22, ferrantraite/IS; 23, I'm Friday/SS; 24, StreetVJ/SS; 24-25, Waymo; 25, Scharfsinn/SS; 26, Ned Snowman/SS; 27, Alexas_Fotos/Pixabay; 28-29, milanzeremski/SS; 29, Gorodenkoff/SS; 30, (UP) Itsuo Inouye/AP/SS; 30, (LO) Ned Snowman/SS; 31, (UP LE) paparazzza/SS; 31, (CTR LE) David Gilder/SS; 31, (LO LE) FeelGoodLuck/SS; 31, (UP RT) esp2k/SS; 31, (CTR RT) Andrey_Popov/SS; 31, (LO RT) Alexas_Fotos/Pixabay; 32, (CTR) Dmytro Zinkevych/SS; 32, (LO) metamorworks/SS; Back cover, (UP) FeelGoodLuck/SS, (LO LE) David Gilder/SS, (LO CTR) Gorodenkoff/SS